The Analysis of Owl Pellets

by D. W. Yalden

GW00600839

Contents

Introduction 3

Dissecting the pellets 4

Key to the identication of mammal skulls 6

Identification of other bones 12

Invertebrate remains 14

Other remains 15

Examples of owl pellet analyses and what we
can learn from them 16

Further analysis 18

Age indicators 18

Sex determination 19

Estimating the size and weight of prey consumed 19

Predicting body weight from linear measurements 21

References 23

Other useful sources of information 25

The Analysis of Owl Pellets, 4th edition
by D. W. Yalden

Published by The Mammal Society

The Mammal Society
Registered Charity No. 278918
Registered Office:
The Mammal Society
3 The Carronades, New Road
Southampton SO14 0AA

ISBN 978-0-906282-67-0

This booklet replaces earlier versions by D.W. Yalden
(2003, 3rd Edition and 1975, Occasional publication No.2)
and by D.W. Yalden and P.A. Morris
(1990, Occasional Publication No. 13)

Drawings by D.W. Yalden
Front cover photograph by Brian Phipps

Layout and printing by SP Press, Units 1 & 2, Mendip Vale Trading Estate,
Cheddar Business Park, Wedmore Road, Cheddar, Somerset BS27 3EL

Introduction

Owls feed mainly on small mammals. They usually swallow their prey whole, and after digesting the flesh, regurgitate the indigestible fur and bones as compact pellets. Because small mammals can be readily identified from their skulls, which are likely to be reasonably intact in the pellets, the analysis of pellets is an extremely valuable tool for the mammalogist. We can learn not only what the owl has eaten, but also which species are present in the locality. Owl pellets are particularly useful for detecting the presence of less common species that are unlikely to be found during small-scale trapping efforts.

Owls usually produce one or two pellets per night and these are most likely to be found at places where the owls roost regularly. In this respect the most "useful" species is the Barn Owl; it roosts in old buildings, cave entrances and hollow trees, and casts at least one of its pellets there each day, gradually accumulating a heap which may contain the remains of hundreds of small mammals. Unfortunately Barn Owls are now rather scarce, and specially protected: it requires a licence to visit an occupied nest site. The much commoner Tawny Owl tends to leave its pellets at a variety of roosts, often evergreen trees, scattered over its territory. Large samples cannot be collected so easily from this species. Short-eared and Long-eared Owls have their favourite roosting places where pellets accumulate. Short-eared Owls roost on the ground, in grassland or moorland, where the pellets are exposed to wind and rain, and soon decompose, but Long-eared Owls choose evergreen tree roosts, like Tawny Owls, and hawthorn thickets. Little Owls eat fewer mammals than their larger cousins, so their pellets are likely to be of more interest to entomologists, and bones may be more broken. Isolated posts and trees may also yield pellets, though often only one or two. Most pellets will contain local prey, but Short-eared and Long-eared Owls are nomadic or migratory, and may bring prey from some distance. Mole remains have been deposited in Ireland by passing owls!

Many other birds also produce pellets, and some of them (notably Buzzards and Kestrels) are also predators of small mammals and interesting to mammalogists. However, birds of prey are stronger than owls, and regularly decapitate their prey or tear it apart before consumption. Their pellets therefore represent a less complete record of their diet. Moreover, their digestion is so thorough that bones may disappear completely (Yalden & Warburton, 1979; Yalden & Yalden, 1985; Andrews, 1990). Isolated teeth may be the only identifiable remains, and the matrix must be searched carefully to find them. Fox droppings are sometimes mistaken for owl pellets, but they are usually foul smelling, and pointed at one or both ends whereas pellets are rounded. Any long bones they contain will have been broken by chewing (Andrews, 1990), and there may be other food remains (e.g. blackberry pips) which do not occur in owl pellets. Small mammal remains may be present, and be identifiable with this key, but often only fur survives, and it is essential to try to identify the mammal remains from their fur to get a full idea of the diet of Foxes (see Reynolds & Aebischer, 1991; Teerink, 1991). Bear in mind that Foxes may scavenge round owl roosts, and mix their droppings with the pellets. Shape and size, particularly breadth, can be used to identify the likely source of pellets, along with roost type and habitat, but seeing the bird or finding shed feathers remain the only certain ways of doing so. Pellet colour does not help with the identity of the predator, but does indicate the type of food e.g. lighter grey for bird feathers.

Table 1.
Identifying pellets.

Species	Roost	Length x Breadth (mm)	Notes
Barn Owl	Holes in buildings, trees, caves	30-70 x 25-30	Shiny, "varnished"
Tawny Owl	Trees	35-85 x 25-30	Thicker, rounder
Long-eared Owl	Evergreen trees	20-75 x 10-25	Thinner, elongate
Short-eared Owl	Ground	20-75 x 10-25	Thinner, elongate
Little Owl	Holes in trees, rocks, burrows	15-25 x 10-15	Thin, bones intact
Kestrel	Holes in trees, cliffs	30-35 x 10-15	Thin, bones eroded
Buzzard	Trees, cliffs	60-70 x 25-30	Thick, bones eroded

Dissecting the pellets

It is best to tackle each pellet separately. Pellets can be treated in batches, but much information may be lost. They can, for instance, be soaked in water as a batch, when most of the fur floats to the surface and the bones sink, conveniently separating the two. However, this loses any information on the matrix, on invertebrate food, and on how much the owl ate on one day.

Air-dry, then weigh and measure each pellet, to get some idea of how much food value it represents. A bird-ringer's Pesola balance and a plastic ruler are adequate. Dissect out the bones from each pellet in turn, using a pair of forceps and a mounted needle. Some workers advocate dampening the pellet first, but I prefer to work dry. Check that all the likely elements (e.g. both jaws to go with a skull) have been discovered. One or more might be missing, perhaps in the next pellet, but usually they are together. Use the needle to remove fur from around the teeth or their sockets.

Do not ignore the matrix. Fur indicates mammal prey, even when no bones or skulls are present. Feathers, which tend to go very dusty, warn you to look for bird bones, often very fragile and easy to overlook. Look out for bird rings, and report them to the British Trust for Ornithology (BTO). The presence of earthworms is generally indicated by sand grains, and the whole pellet may be composed of soil and vegetation if the owl has eaten many worms (as Tawny Owls often do). Alternatively, a lot of vegetation may herald the presence of other invertebrates such as caterpillars and grasshoppers. Lots of small white scales may warn you to look for the identifiable remains of reptiles or fish.

A binocular microscope is extremely useful, particularly for the yellow chaetae (bristles) of earthworms and small jaws of insects. A hand lens (x8 or x10) will suffice for most purposes. In the field, a standard lens detached from a SLR camera, turned back to front, may make an emergency magnifier. Binoculars, turned back to front, will also serve to magnify a small object held close to an eyepiece lens.

Bear in mind that you have the chance to obtain information on a number of different, though related topics. For example:

(a) Small mammals will probably constitute most of the prey. For rarer species, such as Water Shrew, Dormouse and Harvest Mouse, these will be valuable distribution records, certainly at local level and perhaps nationally. Make sure these get reported to the County Mammal Recorder. The spread of the Bank Vole in SW Ireland has been partly monitored using owl pellet samples around the edge of its range (Smal & Fairley, 1984).

(b) You will obtain information on the food of the predator, for which quantitative data are desirable. All prey items, including invertebrates, should be counted or assessed in some way. It is usual to present at least "minimum numbers of prey", these being counts of the highest numbers in some predetermined categories. For mammals, these are usually skulls, left and right jaws, for birds left and right humerus, or numbers of beaks. It may be best to present the results as weights (see pp. 19 - 20). If many invertebrates are taken, this may be the only useful way. By weighing each pellet, and using the weight of pellets of different types, it is possible to assess the contributions of different prey types (Yalden & Yalden, 1985).

(c) You will also be obtaining information on the balance of the small mammal fauna in the area, and perhaps on seasonal or long-term changes. For instance, a change in the ratio of Common to Pygmy Shrews with altitude in the Pennines and in the Scottish Uplands showed that Pygmy Shrews were more numerous on moorlands. Overgrazing, causing a reduced availability of Field Voles, might be shown by a change to more Wood Mice and Common Shrews (see below). Field Voles are less available to Barn Owls in spring but more numerous in late summer, relative to Common Shrews.

So, I recommend weighing each pellet in turn. Pick out the small mammal skulls and lower jaws. If the matrix suggests other prey, or there are no skulls present, pick out the larger long bones and any other identifiable and countable items such as beetle legs or wing cases. If it seems worthwhile (because the contents of a full pellet are not adequately explained by the bits already recognised), check the matrix for earthworm chaetae, caterpillar jaws, fish scales, etc. If there are no skeletal remains, at least score the pellet as "mammal" (fur matrix), "bird" (feather matrix) or "invertebrate" (soil and vegetation). These categories can be used as proportions of "mixed" pellets. This approach is particularly useful when studying Kestrel diets, and allows the proportions of the major prey items to be quantified (but is not so good for the mammalogist who wants to know about the mammals present!). Keep all the bits used for identification - small grip-top polythene bags, one for each prey type, stored inside a larger bag for the whole collection, are very convenient. Stored in this way, they can be used for later comparisons, and can be sent for checking in cases of doubt. A small reference collection, particularly of bird skeletons, may be useful too; road-casualty sparrows, robins and thrushes can be rotted down in a jam-jar and will help enormously to sort out bird bones.

An example of what owl pellet analysis can reveal

Pellets were collected from a Barn Owl roost near London, where the bird hunted over an ungrazed field of tussocky grass, catching mostly Field Voles which are abundant in this habitat type. The landowner later allowed the field to be heavily grazed by horses, which removed the dense grass cover. The owls were able to find fewer of their preferred prey, and had to eat other things which are harder to catch. Soon after the second batch of pellets was collected, the owl disappeared. This was one small part of the general decline of the Barn Owl in recent years, and the pellet analyses helped understand the cause in this case (**Pat Morris**, *pers. comm.*) Using Barn Owl pellets in this way is a valuable tool for monitoring the state of the ecosystem (see p.16).

(Note: scale line on all figures = 1cm approx.)

Key to the identification of mammal skulls

Fig. 1 The skulls of (3) an insectivore and (9) a rodent.

❶ Continuous row of teeth (**insectivores, bats**) go to **3**

3

❷ Gap between incisors and cheek teeth (**rodents**) go to **9**

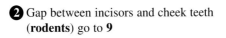

Fig.1

9

❸ Elongate skulls and jaws, lower incisor procumbent (almost horizontal). Teeth usually red-tipped (**shrews**) go to **4**

Note: on the Channel Isles and Isles of Scilly, similar shaped remains but without the red-tipped teeth belong to white-tooted shrews, *Crocidura*

Fig. 2 The lower jaws of
(6) Water Shrew
(5a) Common Shrew
(5b) Pygmy Shrew
(7) Mole and
(8) Bat

OR

Elongate skulls and jaws, much larger and without procumbent lower incisor (**Mole**) go to **7**

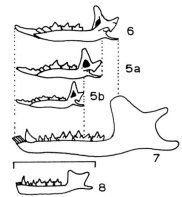

Fig.2

OR

Short squat skulls and jaws, no procumbent lower incisors, big canines, teeth white (**bats**) go to **8**

Fig. 3 Upper & lower
jaws of Common Shrew.

❹ Shrews. Check upper jaw in profile for the number of small single cusped teeth (unicuspids) between hooked first incisor and cheek teeth.

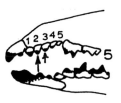

Fig.3

5 unicuspids, lower procumbent incisor with 4 cusps (***Sorex***) go to **5**

OR

4 unicuspids, lower procumbent incisor with smooth blade and one cusp (***Neomys***) go to **6**

❺ *Sorex*. Size larger, 3rd unicuspid smaller than 2nd. Two front cusps of lower incisor clumped **Common Shrew *Sorex araneus*** see **5a** (Common prey of all owls)

OR

Fig. 4 Upper & lower jaws of (5a) Common Shrew (5b) Pygmy Shrew and (6) Water Shrew.

Size smaller, 3rd unicuspid larger than 2nd. Four cusps of lower incisor evenly spaced **Pygmy Shrew *Sorex minutus*** see **5b** (Fairly common as prey, easily missed because of small size)

Fig.4

Fig. 5 Water Shrew skull (6) compared with Common Shrew (5).

❻ *Neomys*. Largest of the shrews. The blade-like lower incisor is distinctive (but beware worn common shrew, which will then have little red pigment remaining). Shape of skull, with concave forehead, also distinctive. **Water Shrew *Neomys fodiens*** (Scarce in owl pellets)

Note. The white-toothed shrews *Crocidura* of the Scilly and Channel Isles have smooth lower incisors like *Neomys,* but they have only 3 unicuspids, and no red enamel on their teeth.

Fig.5

Fig. 6 Mole skull (7) and bat skull (8).

❼ Mole *Talpa europaea*. Skull not common in owl pellets. Looks like enlarged shrew skull, but without the procumbent lower incisors. Usually recognised from humerus - see limb bones, p. 12

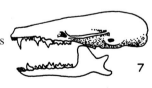

❽ Bats. Very rare in owl pellets. Recognised as bats by short jaws and white teeth, prominent canines. Size varies with species. For further identification, see Yalden 1985a, or Stebbings *et al.* 2007.

Fig.6

Fig. 7 Cheek teeth of vole (10) and mouse (15).

❾ Rodents

With 3 cheek teeth in each jaw, each with zig-zag chewing surfaces (**voles**) go to **10**

OR

With 3 cheek teeth in each jaw, each with small rounded cusps (**rats and mice**) go to **15**

OR

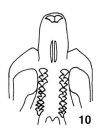

Fig. 8 Cheek teeth (11) and lower jaw (11a) of Dormouse.

With 4 cheek teeth in each jaw. These are flat, with faint, nearly parallel, cross ridges. The lower jaw has a characteristic hole in the angle - **Dormouse *Muscardinus avellanarius*** see **11**
(Note: Dormouse is very rare in owl pellets)

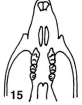

Fig.7

❿ Voles

Fig. 9 Upper (above) & lower (below) molars of Bank Vole (13) and Upper (above) & lower (below) molars of Field Vole (12).

Loops of zig-zags more rounded: second upper molar (m^2) without extra loop on inner (tongue) side, lower molars with loops opposite, especially on second (m^2) and third (m^3) lower molars. Molars develop two roots each with age.
Bank Vole *Myodes glareolus* see **12**
(Note: Fairly common, but much rarer in owl pellets than Field Vole. On Orkney and Guernsey, Bank Voles are absent, but Common Vole *Microtus arvalis* is present. This has teeth essentially like those of *Microtus agrestis* but without the extra loop on m^2)

OR

Loops of zig-zags more angular: m^2 with extra loop on inner side at hind end, lower molars with loops alternate. Molars open-rooted throughout life.
Field Vole *Microtus agrestis* see **13**
(Note: much the most common prey of all owls except Tawny Owl)

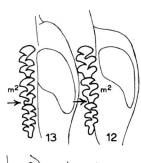

Fig.8

Fig.9

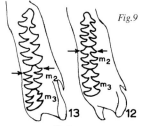

OR

Fig. 10 Lower jaw of Water Vole (14), compared with Field Vole (13).

Much larger than Field Vole, but with similar angular zig-zags, and molars open-rooted throughout life.
Does not have the extra loop on m^2.
Water Vole *Arvicola terrestris* see **14**

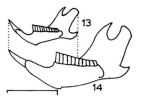

Fig.10

🅑 Rats and Mice

The four genera can be fairly readily distinguished on size, once you have had some practice. The number of root holes left by the first upper molar (m^1) if it is removed, and the shapes of both upper and lower first molars, are distinctive.

Fig. 11 Side view of incisor notch distinguishes Apodemus (16), from Mus (17).

Upper incisor not notched in profile; m^1 with 4 roots; lower molars with 6 roots (because m^3 has 2 roots); m^1 with 3 cusps along front edge.

Fig.11

Wood Mouse *Apodemus sp*. see **16**
(Note: usually the second most common rodent in owl pellets. Distinguishing the two species is difficult, but lower jaws longer than 16 mm are probably *A. flavicollis*. This is much rarer, and usually *A. sylvaticus* can be assumed. See also Fielding 1966, Yalden 1984).

OR

Fig. 12 The shape of lower first molar helps to distinguish the genera. Apodemus (16), Mus (17), Micromys (18), and Rattus (19).

Upper incisor notched in profile; m^1 with 3 roots; lower molars with 5 roots (because m^3 has 1 root); m^1 with 1 asymmetric cusp along front edge.
House Mouse *Mus domesticus* see **17**
(Note: much less common in owl pellets than *Apodemus*. Appreciably smaller, and may mislead into a misidentification as *Micromys*, see next).

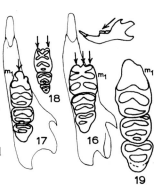

Fig.12

Fig. 13 First upper left molar removed to show differing number of root holes Apodemus (16), Mus (17), Micromys (18), and Rattus (19).

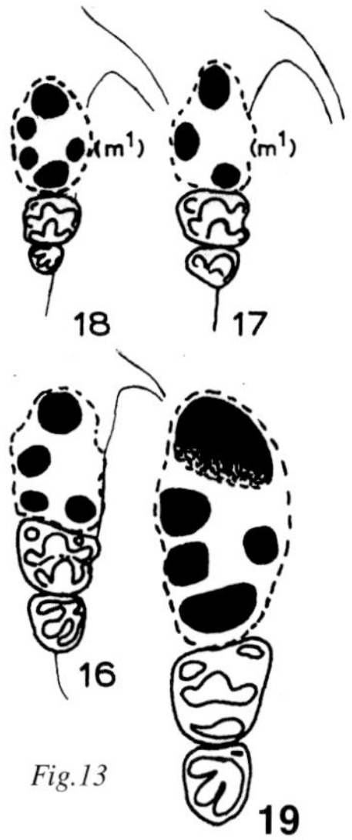

OR

Very small. Incisors not notched in profile; m¹ with 5 roots; lower molars with 7 roots (because m₁ has 3 roots); m₁ with 3 cusps along front edge.
Harvest Mouse *Micromys minutus* see **18**
(Note: Generally uncommon, but may be locally abundant; skulls very fragile, and rarely extricated intact)

OR

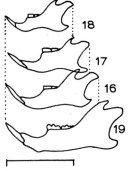

Fig.13

Fig. 14 The lower jaws of Micromys (18), Mus (17), Appodemus (16), and Rattus (19).

Size very large. Incisors not notched; m¹ with 5 roots; lower molars with 10 roots in all (because m₁ has 4 roots, m²⁺³ have 3 roots each); m₁ with 1 cusp on front edge.
Rat *Rattus sp.* see **19**
(Note: Distinguishing the two rat species on incomplete, usually juvenile, skulls is difficult. The Black Rat *Rattus rattus* is so rare that identification as Brown Rat *Rattus norvegicus* can usually be assumed).

㉑ Oddments

Fig.14

Fig. 15 Rabbit teeth (20), Weasel jaw (21) with prominent canine arrowed.

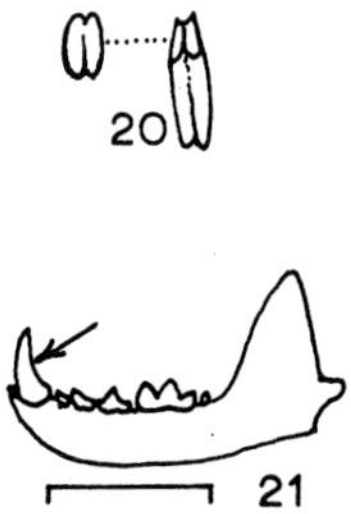

Occasional mammal remains may be encountered that do not fit the key. Very young rabbits may be represented by isolated cheek teeth. These have a crown made of two ovals, and there is a deep groove down each side of the tooth (see **20**). Weasel skulls also occur very occasionally; these are robust, and can be recognised by their large size, large canines and sharp-cusped cheek teeth (see **21**).

Fig.15

Identification of Other Bones

Most limb bones will belong to the small mammals already identified from their skulls. If your interest is simply to identify the local small mammal fauna, there is no need to worry about limb bones. However, if your interest is a full determination of the diet of the owl, then it is important to recognise other components as well. It is especially important to recognise the "ordinary" limb bones, so that you notice the "extraordinary" ones, either larger mammal bones, indicating Water Vole, rats or Rabbit, or bones of very different shapes from frogs, birds, etc.

Fig. 16 Typical mammal bones.

22. Ordinary looking long bones, but much larger than usual. Probably young Water Vole, Rat (**22a**) or Rabbit (**22b**): may be rough at the ends because the epiphyses have dropped off. The rabbit humerus has a distinctive foramen distally (arrowed). There may be no evidence of the skull, because the owl has decapitated its prey, or regurgitated it in a separate pellet.

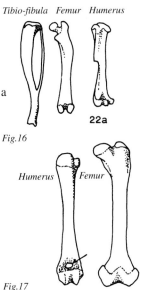

Tibio-fibula Femur Humerus

Fig.16

22a

OR

Fig. 17 Rabbit limb bones.

A long bone, but not of usual mammal type - go to **23**

Humerus   *Femur*

Fig.17

22b

OR

An odd-shaped bone, as broad as long, or not an obvious long bone, perhaps thin sheets of bone - go to **25**

Fig. 18 Typical bones of Frog and Toad.

23. Double-barrelled long bones, evidently the result of two bones joined together, often appearing hollow. **Anura - Frog** or **Toad** (radio-ulna or tibio-fibula) - see **23** Note: quite common in owl pellets. Other bones of the skeleton will occur with these double-barrelled bones, and are also distinctive. The lengths of the bones gives a good indication of the size of prey.

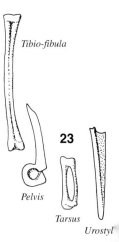

Radio-ulna

Tibio-fibula

23

Pelvis

Tarsus

Urostyl

Fig.18

Fig. 19 Principle long bones of birds.

OR

Humerus Metacarpus Coracoid Metatarsus

Elongate bones, with characteristic triple pulley on one end (metatarsals), with complex depressions and condyles (humerus), or other shapes, all very distinctive.
Bird see 24

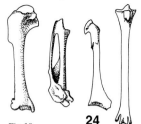

Fig.19 **24**

25. Not an obvious long bone

Fig. 20 Mole humerus.

A sturdy bone, with a peculiar hour-glass shape
Mole *Talpa europaea* (humerus) - see 26

Fig.20 **26**

OR

Thin, rather fragile bones - go to **27**

Fig. 21 Large bones of bird.

27. Thin, almost transparent bone, occurring with feather matrix.
Bird (sternum, sacrum, skull) - see 27

These items are not very useful for determining the size or identity of the bird, being difficult to extract intact, except for the beak, which may be detached from the skull.

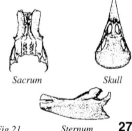

Sacrum   *Skull*

Fig.21   *Sternum* **27**

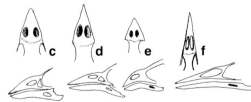

(a) Blackbird
(b) Starling - more slender and with bent lower jaw
(c) House Sparrow
(d) Chaffinch - more slender bill
(e) Yellowhammer - buntings too have a bent lower jaw
(f) Meadow Pipit

OR

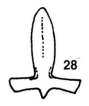

Fig. 22 *Floor of brain case of Frog or Toad (28), fish vertebrae (29).*

Small cross-shaped flat bone
Frog or **Toad** (floor of braincase) - see **28**

Will occur with the double-barrelled long bones (**23**). The skulls of frogs and toads are fragile and fall apart, so you will not recognise them as skulls.

OR

Thin, translucent, bones, parts of the skull of fish; will occur with distinctive spool-like vertebrae, concave at both ends. Their size can be used to estimate the size of the original prey (Wise 1980)
Fish - see **29**

Fig.22

Invertebrate remains

A number of invertebrate remains occur regularly in owl pellets and are illustrated here.

Fig. 23 *Typical beetle fragments, elytra and legs of dung beetle (30a) and ground beetle (30b).*

Shiny purple, oval, leg segments, and other rough leg segments, along with thin wing cases.
Dung beetles, *Geotrupes* - (see **30a**)

Longer thinner legs, and stronger more elongate wing cases, often ridged more strongly than dung beetles, but *Carabus* itself has granular rather than ridged wing cases.
Ground beetles, Carabidae - (see **30b**)

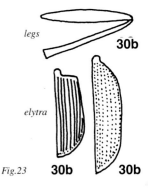

Fig.23

Caterpillars (see **31a**) and **grasshoppers** (see **31b**) are indicated by a matrix of vegetable fragments. Check for minute shiny brown chitinous jaws. Their size can be used to indicate the size of the original prey insects (Yalden & Yalden, 1985). (N.B. scale line here = 1mm).

Earthworms will be indicated by the matrix of the pellet being very gritty. Check some of this grit under a microscope or lens for the minute shiny golden bristles (chaetae - see **31c**) to confirm their presence. Their size indicates the size of their original owner, and their abundance is some measure of the number of worms eaten (see Wroot, 1985).

Fig. 24 Jaws of caterpillars (31a) and grasshoppers (31b) and bristles of earthworms (31c). Scale bars = 1mm.

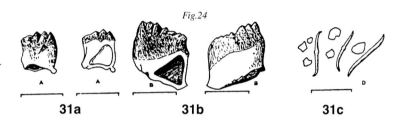

Fig.24

31a **31b** **31c**

Diurnal raptors, particularly Kestrels, take a somewhat different range of miscellaneous items, including lizards, snakes and grasshoppers, some of which are considered by Yalden & Warburton (1979).

Other remains

Fig. 25 Chalky-white tooth plates from house sparrow beak (upper, 31d, 31e; lower, 31f), lizard pelvic girdle (31g) and earwig forceps (31h) are examples of other items sometimes found.

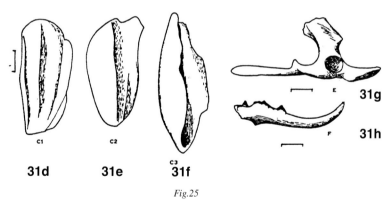

31d **31e** **31f**

31g

31h

Fig.25

Examples of owl pellet analyses and what we can learn from them

The relative abundance of the usual prey mammals in the diets of the different owls is indicated in the following pie charts. These show the numbers of prey (not quite the same as the mass of prey, see later) for the average diets, and may help to indicate that your identifications are correct. For example, if you think you have found that a Barn Owl has been taking 75% Bank Voles, it is evident that this is likely to be an unusual diet. Check that you have applied the relevant part of the key correctly.

Barn owls

Barn Owls normally hunt over rough grassland, along hedges and ditches, and young plantations, where the Field Vole is particularly common. This is their preferred prey, and typically forms 45% of the diet (Fig.26), with Common Shrews and Wood Mice contributing another 25% and 15% respectively. In the absence of Field Voles, they are forced to hunt other prey. A recent survey suggests that they eat fewer Field Voles and Common Shrews, but more Wood Mice and Pygmy Shrews, than they did in the 1960s (Love *et al.*, 2000). Strachan (1995) found that Water Voles declined in the diet of Barn Owls in Co. Durham from 25% in the 1980s to zero in the 1990s, matching the national decline. The study of owl pellets is therefore a useful environmental monitoring technique.

Fig. 26 The diet of the Barn Owl in Britain, (left) in the 1960s, based on 182 batches of pellets from 143 sites containing a total of 46,130 prey items (from Glue, 1974) and (right) from the 1990s, based on pellets from 81 sites containing 48,996 prey items (from Love et al., 2000).

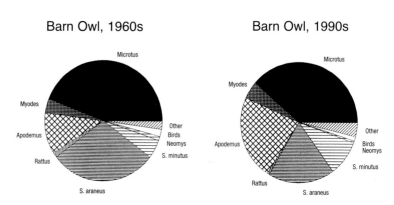

Barn Owl, 1960s

Barn Owl, 1990s

Fig. 26

Long eared owls

Long-eared Owls roost and nest in woodland, but hunt over open ground. They tend to take more birds than other owls normally do and fewer shrews (Fig. 27).

Fig. 27 *The diet of the Long-eared Owl in Britain and Ireland, based on 7,761 prey items from 51 localities (data from Glue & Hammond, 1974).*

Fig. 28 *The diet of the Short-eared Owl in Britain and Ireland, based on 4,120 prey items from 38 localities (data from Glue, 1977).*

Short eared owls

Short-eared Owls live and hunt in open fields and moorland where they feed mainly on Field Voles. They encounter relatively few other species (Fig. 28), though they too may take birds, especially in winter, and take more Pygmy Shrews than the other owls.

The tawny owl

The Tawny Owl is very adaptable. In woodland it takes mostly small mammals, while in urban habitats it switches to eating mainly birds (Fig. 29).

Fig. 29 *The diet of the Tawny Owl in woodland (Bookham Common, Surrey) and urban (Morden, London) habitats (data from Beven, 1965).*

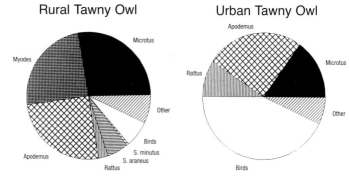

Fig. 29

Analysis of Tawny Owl pellets collected at different times of the year shows a seasonal change in their prey and hunting behaviour. The changes are even more marked when differences in the relative size of prey species are taken into account (see p. 18) and the diet analysis is presented as % of total prey weight taken rather than prey numbers (Fig. 30).

During the winter months the owls preyed on small mammals within the woods (mice, voles and shrews), but once the undergrowth had thickened up the owls had to leave the woods and hunt over open fields. There they took baby Rabbits and many Moles but come the autumn the owls returned to their woodland feeding areas once again. The presence of moles in the diet of owls is especially interesting because it provides clear evidence that these animals spend much more time active on the surface than is generally realised. Many of the Moles were probably young ones dispersing overground from their mother's territory. A similar analysis for Barn Owls is given by Webster (1973).

Fig. 30 Seasonal variation in the diet of the Tawny Owl at Wytham Woods, Oxford, by % weight of prey (data from Southern, 1954).

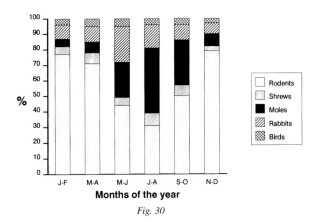

Fig. 30

Further analysis Identifying the numbers of different prey items is only the first step of an analysis though it may be sufficient for most people. However, it is possible to investigate other questions of interest to both the mammalogist and the ornithologist. For example, do the owls prey more heavily on young or old animals? Are males caught more or less often than females?

Age indicators The relative age of mammals is often indicated by features of the skeleton (Morris, 1972) or teeth (Morris, 1978). In shrews, the extensive red tips of young teeth (Fig. 37a) are worn away with increasing age (Fig. 37b). Degree of wear can thus separate age groups (Crowcroft, 1956). In mice the cusps of the molar teeth are worn down with increasing age (Fig. 37c-e; Delany & Davis, 1961).

Fig. 31 Teeth of shrews (37a, 37b) and mice (37c, 37d, 37e) indicating age.

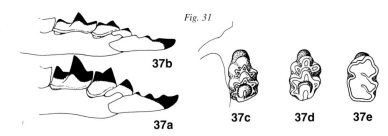

18

In the Bank Vole, the molars develop separate roots and their length gives an indication of age (Fig 38 a-c; Lowe 1971). It is more difficult to assign Field Vole skulls to age classes, but the relative development of the ridges on the skull provides a rough guide; in very young ones, the third molars (both upper and lower) lie in pockets alongside the jaw bone (Fig. 38d), and are absorbed as the jaws elongate with age (Fig 38e).

Fig. 32 Teeth of Bank Voles (38a, 38b, 38c) and Field Voles (38d, 38e) indicating age.

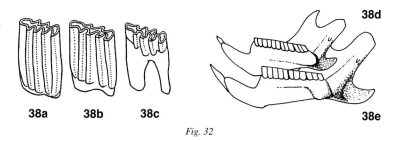

Fig. 32

Sex determination

For most rodents and shrews, the sex can be determined, and immatures separated from adults, by the shape of the pelvis (Dolgov, 1961; Brown & Twigg, 1969).

Fig. 33 The shape of the pelvis in (a) female and (b) male rodents; and in (c) female and (d) male shrews.

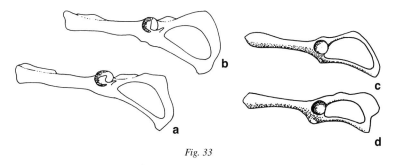

Fig. 33

Estimating the size of prey and total weight consumed

Merely counting numbers of different prey items present in the pellets does not necessarily indicate their relative importance as food: a Harvest Mouse clearly has less meat on it than a Brown Rat. To assess the relative significance of the different prey species we need to consider size differences. Numbers of prey items should be converted into weight of food (biomass) eaten. This involves some degree of approximation, and it is important to bear in mind that this might induce error. If all prey are about the same size or frequency, the effect of error is less than if some are much larger or more abundant.

19

One much-used device has been the "prey unit" based on a standard vole or mouse assumed to weigh 20 grams. Thus, voles and mice are assessed at 1 prey unit (p.u.) each, and other species scaled accordingly, e.g. Common Shrews are scored as 0.5 p.u. on the assumption they weigh half as much as a mouse, about 10g (Table 2). This system was proposed by Southern (1954) and has been used extensively since.

Table 2. Prey unit conversion factors (multiply by numbers of prey items found to estimate biomass in terms of prey units).

Apodemus, Mus, Microtus, Myodes	= 1.0 p.u.
Micromys, Sorex araneus	= 0.5 p.u.
Sorex minutus	= 0.3 p.u.
Neomys	= 0.75 p.u.
Rattus, Arvicola, Talpa, Oryctolagus	= 5.0 p.u.

Obviously not all members of a species weigh the same, nor is 0.5 p.u. (=10g) a realistic estimate of the average weight of a common shrew. Direct estimates of prey weights would be better. Those in Table 3 are biased towards the weights of younger animals which are more numerous in the population.

Table 3. Realistic average weights (g) for potential prey species of owls

Sorex araneus	8	*Muscardinus avellanarius*	15
Sorex minutus	4	*Micromys minutus*	5
Mus domesticus	12	*Apodemus flavicollis*	25
Apodemus sylvaticus	18	*Myodes glareolus*	16
Neomys fodiens	12	*Microtus agrestis*	21
Talpa europaea	70	*Arvicola terrestris*	80
		Rattus sp.	60

Errors creep in if some prey are only partly eaten, which is especially likely for the larger species.

Predicting likely body weight from linear measurements

In practice, small mammals only vary in weight by a few grams, but larger species are more variable; rats may weigh anything from 20 g (when they leave the nest) to 700 g as old adults. If a lot of rats are included in the diet, their importance relative to other species depends upon what they are all assumed to weigh.

Since growth in length is linked to growth in body size, weight can be individually predicted from measurement of the length of the lower jaw (the part of a rat most frequently found intact in owl pellets) as shown in Fig. 40.

An estimate of body weight can be made by reading off the graph of lower jaw lengths against actual body weights (Fig. 41). Using this method to estimate the weights of 69 rats actually eaten by Barn Owls showed that they ranged from 1 to 8 prey units (26-164 g), averaging 60 g (Morris 1979).

Fig. 34 *Measurement of the length of a rat lower jaw.*

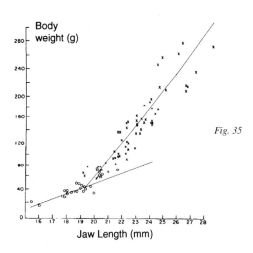

Fig. 34

Fig. 35 *The relationship between jaw length and body weight in laboratory-reared rats (Morris, 1979). The lower line is for very young animals with their third molars not yet fully erupted.*

Fig. 35

Alternatively, likely body weight can be estimated from the formula:
$$\log weight = (4.7170 \times \log jaw\ length) - 4.2923$$

21

Although birds are difficult to identify when they occur in owl pellets, the length of their humerus (in millimetres) (see p. 12) gives a good indication of their original weight (in grams). The relative importance of birds in the diet can therefore be assessed even when their identity is uncertain. The graph here (Fig. 42) indicates the relationship for the size range which is most useful. Other bone measurements can also be used but are less reliable (see Morris & Burgis, 1988). Alternatively, weight can be predicted from the formula:

$$\log \text{weight} = (2.4221 \times \log \text{humerus length}) - 3.8027$$

Fig. 36 Relationship between humerus length and body weight in passerine birds (from Morris & Burgis, 1988).

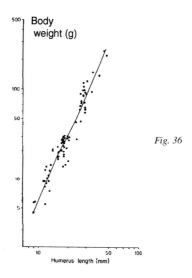

Fig. 36

Frogs vary in size, and males are smaller than females. Mature adults weigh 20-60 g; a reasonable average would be about 45 g.

Estimating the contributions by weight of other prey is very difficult. Beetles vary in size too, from 0.3 to 1.4 g in *Geotrupes*. Fish sizes can be estimated from vertebrae (Wise, 1980) but are unlikely to form a major part of any owl's diet.

The contribution made by earthworms can be estimated by counting and measuring chaetae (Wroot, 1985), or by assuming that 1 g of "sand" in a gritty pellet comes from 10 g of earthworms (Yalden & Warburton, 1979). In order to do this you need to weigh each pellet before you analyse it. If it has a weight of 1 g, and a matrix of 50% fur and 50% sand, assume that 50% of the 1 g came from earthworm prey. At the end of the batch of pellets, total the weight of pellets that came from earthworms and multiply by 10 to get an estimate of their weight, which you can then compare with the weight of mammals (and other prey) which were eaten. While there are some obvious sources of error in this procedure (sand is more dense than fur!), the errors are less than would result from ignoring invertebrate prey completely. Tawny Owls in particular (also Kestrels) may take up to 20% of their diet as invertebrates (Yalden, 1985b).

References

Andrews, P. (1990) *Owls, Caves and Fossils*. British Museum (Natural History), London.

Beven, G. (1965) The food of tawny owls in London. *London Bird Report*, **29**, 56-62.

Brown, J.C. & Twigg, G.I. (1969) Studies on the pelvis in British Muridae and Cricetidae (Rodentia). *Journal of Zoology, London*, **158**, 81-132.

Crowcroft, W.P. (1956) On the life span of the common shrew (*Sorex araneus* L.). *Proceedings of the Zoological Society of London*, **127**, 286-292.

Delany, M.J. & Davis, P.E. (1961) Observations on the ecology and life history of the Fair Isle field mouse *Apodemus sylvaticus fridariensis* (Kinnear). *Proceedings of the Zoological Society of London*, **136**, 439-452.

Dolgov, V.A. (1961) Variation in some bones of postcranial skeletons of the shrews (Mammalia, Soricidae). *Acta Theriologica*, **5**, 203-227 (in Russian, but well illustrated, and summarised in English).

Fielding, D.C. (1966) The identification of skulls of the two British species of *Apodemus*. *Journal of Zoology, London*, **150**, 491-511.

Glue, D.E. (1974) Food of the barn owl in Britain and Ireland. *Bird Study*, **21**, 200-210.

Glue, D.E. (1977) Feeding ecology of the short eared owl in Britain and Ireland. *Bird Study*, **24**, 70-78.

Glue, D.E. & Hammond, G.J. (1974) Feeding ecology of the long eared owl in Britain and Ireland. *British Birds*, **67**, 361-369.

Love, R.A., Webbon, C., Glue, D.E. & Harris, S. (2000) Changes in the food of British Barn Owls (*Tyto alba*) between 1974 and 1997. *Mammal Review*, **30**, 107-129.

Lowe, V.P.W. (1971) Root development of molar teeth in the bank vole (*Clethrionomys glareolus*). *Journal of Animal Ecology*, **40**, 49-62.

Morris, P.A. (1972) A review of mammalian age determination methods. *Mammal Review*, **2**, 69-104.

Morris, P.A. (1978) The use of teeth for estimating the age of wild mammals. In: *Development, Function and Evolution of Teeth* (Ed. by K.A. Joysey), pp. 483-494. Academic Press, London.

Morris, P. (1979) Rats in the diet of the Barn Owl (*Tyto alba*). *Journal of Zoology, London*, **189**, 540-545.

Morris, P.A. & Burgis, M.J. (1988) A method for estimating total body weight of avian prey items in the diet of Owls. *Bird Study*, **35**, 147-152.

Reynolds, J.C. & Aebischer, N.J. (1991) Comparison and quantification of carnivore diet by faecal analysis: a critique, with recommendations, based on a study of the Fox *Vulpes vulpes*. *Mammal Review*, **21**, 97-122.

Smal, C.M. & Fairley, J.S. (1984) The spread of the bank vole *Clethrionomys glareolus* in Ireland. *Mammal Review*, **14**, 71-78.

Southern, H.N. (1954) Tawny Owls and their prey. *Ibis*, **96**, 384-410.

Stebbings, R.E., Yalden, D.W. & Herman, J.S. (2007) *Which Bat Is It?* The Mammal Sociey, London.

Strachan, R. (1995) *Mammal Detective*. Whittet Books, London.

Teerink, B.J. (1991) *Hair of West-European Mammals*. Cambridge University Press, Cambridge.

Webster, J.A. (1973) Seasonal variation in mammal contents of Barn Owl castings. *Bird Study*, **20**, 185-196.

Wise, M.H. (1980) The use of fish vertebrae in scats for estimating prey size of otters and mink. *Journal of Zoology, London*, **192**, 25-31.

Wroot, A.J. (1985) A quantitative method for estimating the amount of earthworm (*Lumbricus terrestris*) in animal diets. *Oikos*, **44**, 239-242.

Yalden, D.W. (1984) The yellow-necked mouse, *Apodemus flavicollis*, in Roman Manchester, *Journal of Zoology, London*, **203**, 285-288.

Yalden, D.W. (1985a) *The Identification of British Bats*. Occasional Publication No. 5, The Mammal Society, London.

Yalden, D.W. (1985b) Dietary separation of owls in the Peak District. *Bird Study*, **32**, 122-131.

Yalden, D.W. & Warburton, A.B. (1979) The diet of the kestrel in the Lake District. *Bird Study*, **26**, 163-170.

Yalden, D.W. & Yalden, P.E. (1985) An experimental investigation of examining Kestrel diet by pellet analysis. *Bird Study*, **32**, 50-55.

Other useful sources of information

Corbet, G.B. (1975) *Finding and Identifying Mammals in Britain*. Natural History Museum, London. (Obtainable through HMSO. Has keys to all mammals, but based on entire skulls or specimens).

Harris, S. & Yalden, D.W. (2008) Mammals of the British Isles : Handbook 4th Edition. The Mammal Society, Southampton.

Lawrence, M.J. & Brown, R.W. (1973) *Mammals in Britain: their Tracks, Trails and Signs* (2nd edn.), Blandford Press, London. (Additional drawings of skulls, bones etc).

Sargent, G. & Morris, P. (2003) *How to Find and Identify Mammals*. The Mammal Society, London.

Useful addresses

The Mammal Society, 3 The Carronades, New Road,
Southampton SO14 0AA
T: 0238 0237 874 F: 0238 0634 726
W: www.mammal.org.uk E: enquiries@mammal.org.uk

British Trust for Ornithology (BTO), The Nunnery, Thetford,
Norfolk IP24 2PU.
T: 01842 750050 F: 01842 750030
W: www.bto.org E: info@bto.org

The Mammal Society's Practical Guides to Mammalogy

Other books in the series include:

Identification of Arthropod Fragments in Bat Droppings

By Caroline Shiel, Catherine McAney, Claire Sullivan and James Fairley

Instruction on the collection of bat droppings and how to extract identifiable remains of insects, arachnids and other arthropods from them.

A Guide to the Identification of Prey Remains in Otter Spraint

By J.W.H. Conroy, J. Watt, J.B. Webb and A. Jones

Includes methods of determining otter diet, collection and storage of spraints and keys to identify fish vertebrae.

How to Find and Identify Mammals

By Gillie Sargent and Pat Morris

An illustrated guide to distinguishing mammals in the field and identifying them by their signs. The manual also includes a small mammal skull key and standard recording forms.

Live Trapping Small Mammals

By J. Gurnell and J.R. Flowerdew

An illustrated guide on how to use small mammal traps, plan your survey and analyse your data.

Which Bat is it?

A Guide to bat identification in Britain and Ireland. Includes identification keys not only of Bats in the hand but also from skulls and teeth. Information is also included on the identification of droppings and flight characteristics.

The Mammal Society also produces a wide range of Books on individual mammal species which include:

- The Otter
- The Pine Marten
- The Dormouse
- The Wildcat
- Stoats and Weasels
- The Water Shrew Handbook
- The Edible Dormouse
- Fallow Deer
- Sika Deer
- Chinese Water Deer

To order any of these books ring The Mammal Society on **0238 0237 874** or order online at **www.mammal.org.uk**

The Mammal Society - the voice for British mammals, and the only organisation solely dedicated to the study and conservation of all British Mammals

Getting involved - how you can help
The Mammal Society

Britain is home to 60 kinds of wild mammals - found on land, in the air, rivers, streams and lakes - plus another 18 seen regularly in the seas around us.

They are some of the most beautiful on Earth. They are also important indicators of the health of our environment.

But at least half these species are under serious threat, The Mammal Society is working to change this - and with your help we can:

Join us

Membership costs from as little as £25 a year. As well as the knowledge that you are helping to safeguard the future of British wildlife you will receive our newsletter *Mammal News* and have the opportunity to take part in a wide range of events from field trips to training weekends and seminars to conferences. There is also the opportunity to receive our scientific journal *Mammal Review*.

Take Part

The Mammal Society relies on a committed band of volunteers to take part in mammal recording and survey work. It is only with YOUR help that we can continue with our survey and monitoring projects and know how to conserve British mammals for the future.

To find out how to get involved with The Mammal Society's latest surveys please e-mail **enquiries@mammal.org.uk.**

You can also join your Local Mammal Group and take part in local mammal recording and surveying as well as participating in a wide range of talks and events.

Make a donation!

The Mammal Society is reliant on donations for a lot of its income. There are many ways to donate and any donation large or small will be of great help to our work.

For details of your Local Mammal Group and County Mammal Recorder or to join The Mammal Society or make a donation visit **www.mammal.org.uk** or phone **0238 0237 874**.